AHA! ACADEMY

SCORCHING SCIENCE!

The Physics of **Heat**

Written by Yvette LaPierre

WORLD BOOK

www.worldbook.com

Co-published by agreement between Shi Tu Hui and World Book, Inc.

Shi Tu Hui
Room 1807, Block 1,
#3 West Dawang Road
Chaoyang District, Beijing 100025
P.R. China

World Book, Inc.
180 North LaSalle Street
Suite 900
Chicago, Illinois 60601
USA

Library of Congress Control Number: 2024947136

Aha! Academy: Physics
ISBN: 978-0-7166-7144-2 (set, hard cover)

Scorching Science! The Physics of Heat
ISBN: 978-0-7166-7153-4 (hard cover)
ISBN: 978-0-7166-7173-2 (e-book)
ISBN: 978-0-7166-7163-3 (soft cover)

Printed in India by Replika Press PVT LTD, Haryana, India
1st printing January 2025

Staff

Editorial

Vice President
Tom Evans

Editorial Project Coordinator
Kaile Kilner

Senior Curriculum Designer
Caroline Davidson

Proofreader
Nathalie Strassheim

Graphics and Design

Senior Visual
Communications Designer
Melanie Bender

Designer
Shannon Hagman

Digital Asset Specialist
Rosalia Bledsoe

Written by Yvette LaPierre
Advised by Dr. Nathalie Haurberg

Developed with World Book by
Red Line Editorial

Acknowledgments

The publishers gratefully acknowledge the following sources for photography. All illustrations were prepared by WORLD BOOK unless otherwise noted.

Cover: Art Stocker/Shutterstock; MIKISTOCKER/Shutterstock; New Africa/Shutterstock; Nitr/Shutterstock; schanz/Shutterstock; Wirestock Creators/Shutterstock

Cavan Images/Alamy Images 4, 11; ESA/Hubble/NASA 36; IBM 39; NASA 43; NASA/Lockheed Martin Corporation/R. Underwood 37; NASA/Tony Gray 37; NRAO/AUI/NSF/NASA/STScI/JPL-Caltech 36, 48; Public Domain 6; Shutterstock 3, 4, 5, 6, 7, 8, 9, 10, 11, 12, 13, 14, 15, 16, 17, 18, 19, 20, 21, 22, 23, 24, 25, 26, 27, 28, 29, 30, 31, 32, 33, 34, 35, 36, 38, 39, 40, 41, 42, 43, 44, 45, 46, 47, 48

There is a glossary of terms on page 48. Terms defined in the glossary are in type that looks like ***this*** on their first appearance on any spread (two facing pages).

Contents

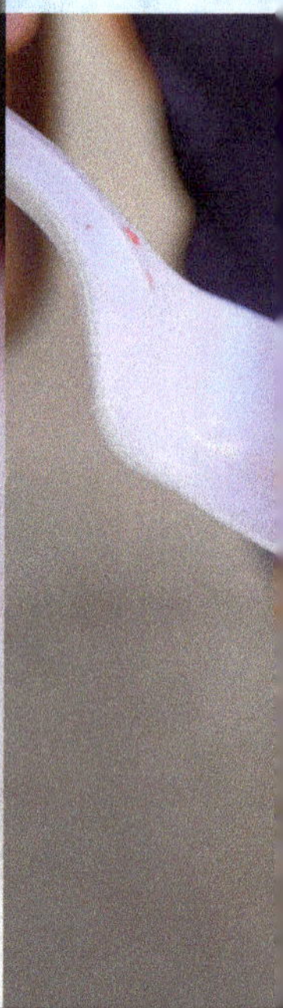

Introduction

You cool off with a frosty snow cone on a sunny day. You warm up with a hot cup of tea on a cold evening. We experience hot and cold every day!

Most people use the word *heat* to describe something that feels warm. But heat is not the same thing as temperature. You'll soon know the difference!

Scientists study the physics of heat to understand how it is created, how it moves, and how it affects materials and the world around us. Read on to learn what's so cool about heat!

Believe it or not, heat transfer is taking place in this glass of iced tea!

1

WHAT IS HEAT?

A volcano erupting with hot lava. A bluish-white iceberg in the chilly ocean. What do these two natural formations have in common? They both contain something called thermal energy!

Thermal energy comes from the movement of tiny particles in an object. Because everything is made up of these moving particles, *everything* contains thermal energy—even you! The faster the particles are moving, the more thermal energy the object contains. The particles in the volcano are moving a lot faster than those in the iceberg, so the volcano has more thermal energy.

DID YOU KNOW?

Thermal energy from deep inside Earth escapes through volcanoes at the surface. The heat released when a volcano called Mount St. Helens erupted in 1980 could have powered the entire world for more than two days!

Physicists define heat as the transfer of energy between two things. Energy flows from warmer objects to cooler ones. So if that volcano and iceberg were in the same room, thermal energy would move from the volcano to the iceberg.

But what is temperature, and how is it different from heat? While heat is the transfer of energy, temperature is a measure of energy. You can see this measure of energy when you look at a thermometer or a thermostat!

Getting kinetic

All *matter* is made up of tiny particles called *atoms* and *molecules*. These particles are in constant motion. They are always vibrating, jiggling, and bouncing around! Due to this motion, particles are said to have kinetic energy. The faster a particle moves around, the more kinetic energy it has.

The total kinetic energy of all the particles in an object is called its thermal energy. This radiator has more thermal energy than this snowperson. Why? Because the particles in the radiator are moving faster, giving it a greater total kinetic energy.

When you wrap your hands around a mug of tea and feel its warmth, you're actually sensing the vibration of tiny particles. Scientists call this kinetic energy. Let's explore it!

Temperature is a measure of the average kinetic energy for each particle in an object. Does that mean an object's average kinetic energy (temperature) increases when its total kinetic energy (thermal energy) increases? Not always! Imagine you have a mug of water with a temperature of 100 °F (38 °C). Now expand that mug of water into a big tub of water. The two containers of water have the same average kinetic energy (temperature). But the big tub has more total kinetic energy (thermal energy) simply because it has more particles!

CAREER CORNER

Thermodynamics is the study of the physics of heat and how different materials behave at different temperatures. Many thermodynamic engineers work in aerospace. For example, a thermodynamic engineer might help build rockets, which are propelled by hot gases.

Turn up the **heat**

The sun is about 93 million miles (150 million kilometers) away from Earth. But it's our most important heat source. Without the sun, our planet would be frozen solid!

Heat is constantly flowing from Earth's core, which is as hot as the surface of the sun. This core heat rises to Earth's surface in the form of volcanoes and hot springs.

DID YOU KNOW?

When you feel cold, you can create heat by rubbing your hands together. The ***friction*** causes the ***molecules*** in your skin to move faster and bump into each other, producing more thermal energy.

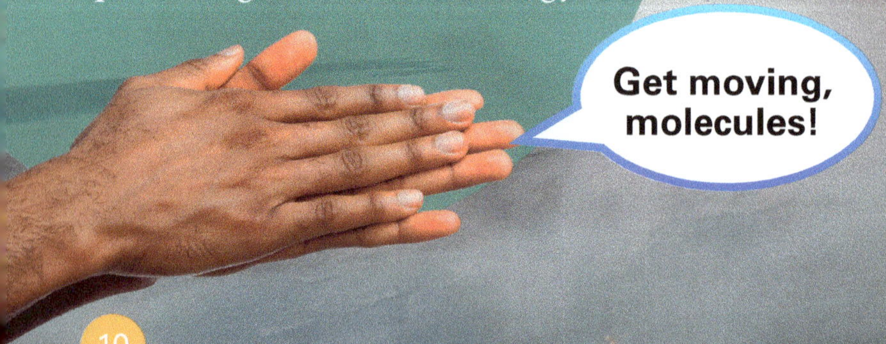

Think of all the ways you can warm up if you're cold. You might sit by a fire, stand in a sunny spot, or even soak in a hot spring. Anything that creates heat is called a heat source. Heat sources can be natural or made by humans.

Many home appliances, such as toasters and stoves, are sources of heat. Electric appliances give off heat by converting electrical energy into thermal energy.

Burning such materials as wood, oil, and gas creates a heat source. This is because burning is a chemical reaction that releases heat.

Even you are a source of heat! When you wrap up in a blanket, the blanket itself isn't warming you up. It is trapping your own body heat against you.

Heat on the **move**

Remember how heat flows from warmer objects to cooler ones? In your glass of ice water, heat flows from the room-temperature water to the ice cubes. This means the water loses thermal energy while the ice cubes gain thermal energy. In other words, the water cools down and the ice warms up! Let's look at some other examples of heat as thermal energy transfer.

Wrap your hands around a mug of hot cocoa on a cold day, and your hands will warm up. That's because thermal energy flows from the hot mug to your cool hands. As your hands warm up, the mug cools down.

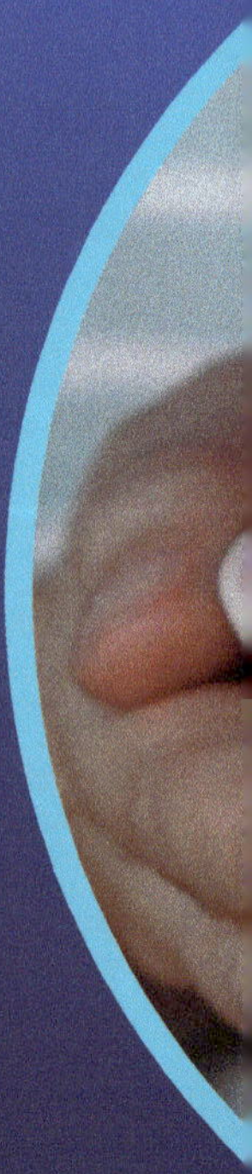

You put ice cubes in a glass of room-temperature water. Five minutes later, your water is nice and cold. Did the ice cubes transfer their cold to the water? No! In fact, your water is cold because it transferred heat to the ice cubes!

The opposite happens when you hold a bowl of ice cream. Now your hands are warmer than the cold bowl, so thermal energy flows from your hands to the bowl. That's why your hands feel cold—because energy is leaving them!

Let's revisit that glass of ice water. After an hour, all the ice cubes are melted. Heat transferred from the water to the ice until the two substances reached the same temperature. This balance in temperature is called ***thermal equilibrium***. It's the point at which heat transfer stops.

Early scientists, such as **Isaac Newton,** studied heat. They thought it was a type of fluid that they named "caloric." Scientists still talk about heat as flowing like a liquid, and they measure it in units called calories!

2 GOING WITH THE FLOW

Heat can flow from one substance to another by three different methods: conduction, convection, and radiation. When you warm up a pot of water on the stove, all three methods of heat transfer are at work!

Conduction is the transfer of heat between two objects or substances in direct contact. The bottom of the pot is touching the hot burner. Thermal energy moves from the burner to the bottom of the pot. Then that thermal energy warms the water in contact with the hot bottom of the pot.

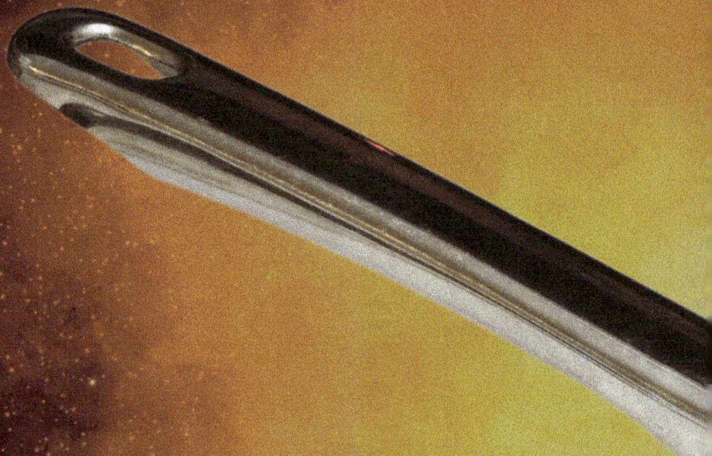

DID YOU KNOW?

On a hot day, a white T-shirt keeps you cooler than a black T-shirt does. That's because the white shirt ***absorbs*** less energy from the sun. It also ***reflects*** more heat away from you.

Convection is the transfer of heat by the total motion of particles within a liquid or gas. The warmer water near the burner rises to the top of the pot. As it travels, it transfers its thermal energy to cooler water in the pot. This process continues until the water is boiling.

Radiation is the transfer of thermal energy in the form of light waves. Heat from the burner travels through the air around it due to radiation.

Conduction

Conduction relies on direct contact for thermal energy to move from one object to another. But how exactly does the energy transfer between the objects? Let's take a closer look at the bowl of soup!

The particles in the hot soup jiggle around faster than those in the metal spoon. In fact, the soup particles have so much kinetic energy that they bounce against the particles in the spoon.

Have you ever left a metal spoon in a bowl of hot soup? After a while, the handle of the spoon warms up. Thermal energy is transferred from the soup to the spoon through conduction!

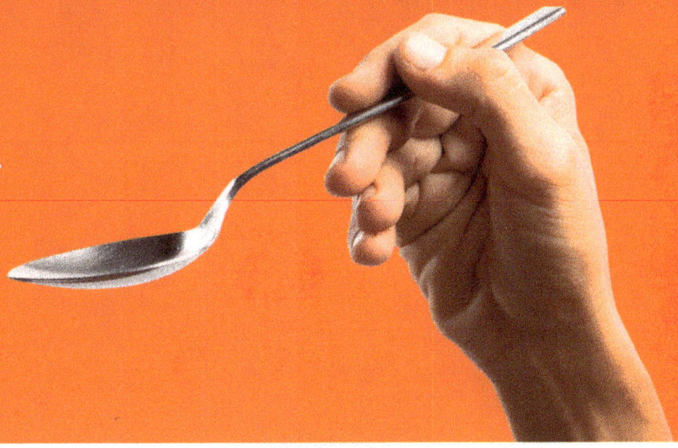

These collisions knock the spoon's particles into faster motion. The faster-moving particles then bump into neighboring particles.

CURIOUS CONNECTIONS

BIOLOGY

Some animals make use of conduction to stay warm! Lizards, for example, are cold-blooded. This means a lizard's body temperature depends on the temperature of its environment. Lizards often lie on warm rocks and let heat transfer from the rocks to their bodies.

As the kinetic energy of the spoon's particles increases, the temperature of the spoon rises!

Convection

Convection is the transfer of heat within a liquid or gas. In this case, the gas is the air inside the balloon. A heater warms this air. As the air near the heater gains thermal energy, its particles move faster and spread farther apart. This particle motion makes the air in the balloon less ***dense***, or heavy—so it rises!

The air at the top of the balloon is cooler because it has less thermal energy. That means the particles at the top of the balloon have less kinetic energy—they move more slowly and stay closer together. This makes the cool air denser, causing it to sink.

When the sinking cool air reaches the bottom of the balloon, it gets warmed by the heater and continues the cycle of rising and sinking. Eventually, the air inside the balloon is less dense than the air outside the balloon, causing the balloon itself to rise into the sky!

Have you ever seen a hot-air balloon rise into the sky? This is possible because of the method of heat transfer known as convection!

Have you ever noticed that the top floor of a building is warmer than the bottom floor? That's because warm air rises within a building just as it rises within a hot-air balloon. As the warm air rises, cool air sinks.

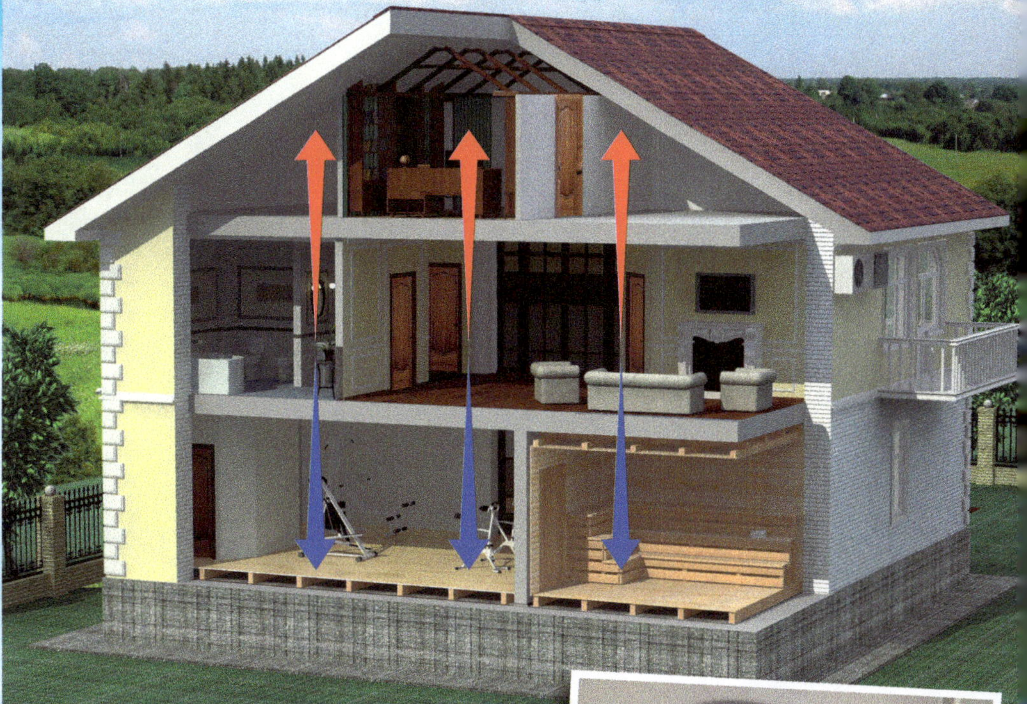

CURIOUS CONNECTIONS

AVIATION

Have you ever been traveling on an airplane and the ride got a little bumpy? These bumps are called turbulence, and they are caused by convection! Warm air rises from Earth's surface into the atmosphere. Meanwhile, cooler air sinks. Planes bump through these areas of moving air.

Radiation

Radiation is the transfer of thermal energy that travels in light waves. These light waves can also travel through outer space, where there is no air!

A fire *radiates*, or sends out, energy in all directions. This heat travels through the air in light waves that reach the cat.

The back of the cat warms up because it's receiving the thermal energy radiating from the fire.

A sleepy cat curls up by a fireplace and quickly falls asleep. It can enjoy the warmth of the fire thanks to the last method of heat transfer: radiation!

The cat radiates heat, too. All ***matter***—including you!—is constantly radiating heat.

TECH TIME

A microwave oven heats food using radiation. Invisible light waves called microwaves travel to the food and cause water ***molecules*** in the food to vibrate. This kinetic energy creates heat that cooks the food.

Wonderful **weather**

The sun heats our planet through radiation. Light waves travel from the sun to Earth through space. Earth's surface ***absorbs*** these light waves and warms up.

DID YOU KNOW?

Convection occurs in the ocean, too! Water near Earth's equator receives more heat from the sun than water near Earth's poles. The cold water at the poles is ***denser***, so it sinks. Warmer water flows in from the equator to take its place. This cycle creates ocean currents that span the entire globe!

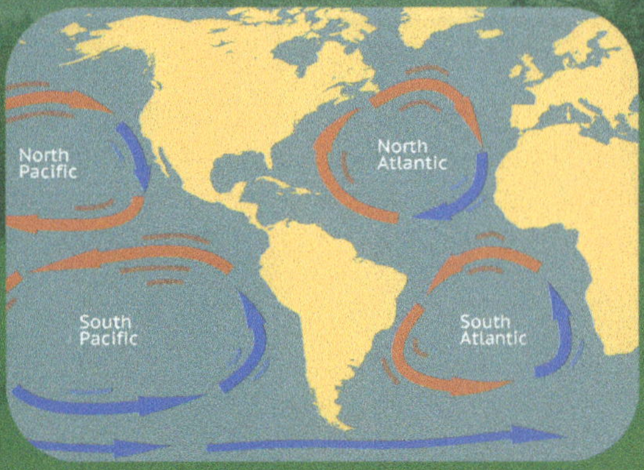

Heat flows to and around Earth through radiation, conduction, and convection. We experience this transfer of thermal energy as weather!

Convection takes place in Earth's atmosphere. The warm air at Earth's surface rises because it is less dense. This rising air causes clouds to form. Meanwhile, cooler, denser air sinks and takes the place of the warm air. We feel this movement of air as wind. Strong winds bring changes in weather!

Earth's sun-warmed surface warms the air above it through conduction.

3

DEGREES OF HEAT

How hot is it outside today? Is the meat fully cooked? Do I have a fever? You can answer all these questions with the help of a thermometer. This is a device that measures temperature.

Temperature is measured using a few different scales. The two most common scales are the Celsius and Fahrenheit scales. Many standard thermometers show temperatures on both scales. A third common scale, the Kelvin scale, is mostly used by scientists.

In simple terms, temperature measures how hot (or not hot) something is. For example, a cook measures the temperature of boiling sugar to make sure it's hot enough to make candy. But let's zoom in on that boiling sugar until we can see the motion of its particles. The temperature of the sugar is actually a measure of how fast, on average, its particles are moving.

The Galileo thermometer is a glass tube filled with clear liquid and glass balls of different ***densities***. The device is named after Italian physicist **Galileo**, who established around 1600 that the density of a liquid changes with temperature. You can observe this in the Galileo thermometer.

As the temperature of the liquid in a Galileo thermometer falls, it becomes denser than the glass balls—so the balls float to the top of the tube. The opposite happens when the temperature of the liquid rises!

Temperature **scales**

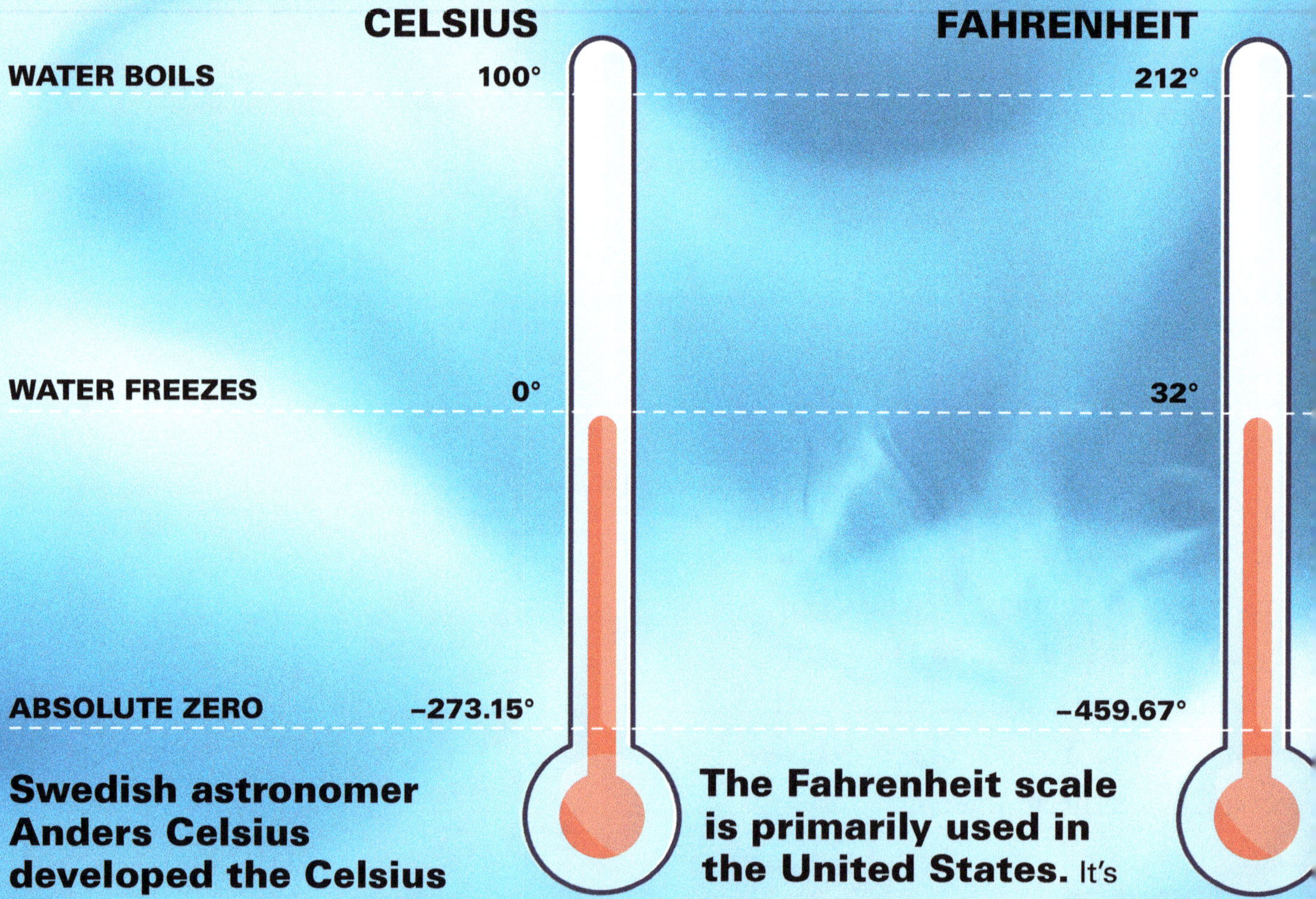

Swedish astronomer Anders Celsius developed the Celsius scale in 1742. It uses units called degrees (°) and is based on the boiling and freezing points of water. Most people around the world use the Celsius scale.

The Fahrenheit scale is primarily used in the United States. It's named after the Dutch physicist Daniel Gabriel Fahrenheit, who developed the scale in 1724. Like Celsius, the Fahrenheit scale uses units called degrees (°) and is based on the boiling and freezing points of water.

To convert a temperature from Celsius (°C) to Fahrenheit (°F), use the formula °F = (°C × 1.8) + 32.

To convert a temperature from Fahrenheit (°F) to Celsius (°C), use the formula °C = (°F – 32) × 0.55.

Celsius, Fahrenheit, Kelvin . . . why are there different scales for measuring temperature?

What are the differences between them all? Let's find out!

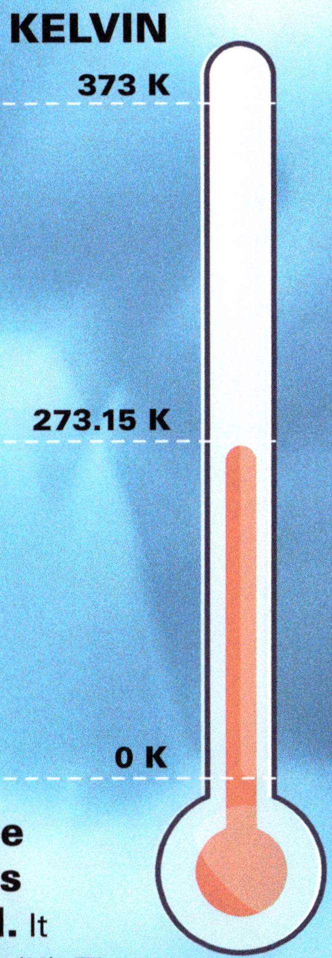

William Thomson was a Scottish scientist who was later given the title Lord Kelvin. In 1848, Thomson developed an absolute zero scale of temperature that became known as the Kelvin scale.

Kelvin is the scale used by scientists around the world. It uses units called kelvins (K). The lowest point of the Kelvin scale is absolute zero (you'll learn more about absolute zero later in the book!). Because of this, the scale doesn't include negative numbers.

To convert a temperature from Kelvin (K) to Celsius (°C), use the formula °C = K – 273.15.

Holding **heat**

When enough heat transfers to an object, the object's temperature increases. But some objects need more heat than others for their temperature to change. These objects are said to have a higher heat capacity.

Let's compare two different foods: a potato and a slice of bread. The potato has more ***mass*** than the slice of bread. Because of its higher mass, it can hold more thermal energy.

The potato also has a different chemical makeup than the bread. For example, the potato contains more water. And water has what's called a high specific heat. Specific heat is the amount of heat required to raise the temperature of a specific amount of a substance by one degree.

Have you ever picked up a baked potato and quickly dropped it because it was too hot?

Baked potatoes stay hot for a while because of something called heat capacity.

Because of the potato's mass and chemical makeup, it has a higher heat capacity than the bread. That is, it needs more heat than the bread for its temperature to increase by one degree. This also means it will take longer for the potato's temperature to decrease by the same amount!

DID YOU KNOW?

Water has a higher heat capacity than sand. You can observe this when you go to the beach on a sunny day. It takes more heat to raise the temperature of the water than it takes to raise the temperature of the sand. That's why the water feels cool while the sand feels hot!

4

HEAT AT WORK

On a hot day, you watch the liquid in your thermometer rise. You see the ice slowly melt in your glass of lemonade. You're observing two important ways that heat affects *matter*. Let's find out what they are!

As matter is warmed, its particles gain kinetic energy—they move faster and farther apart. This increased particle motion causes the matter to expand. Think of people at a party. When they stand close together talking, they take up less space than when they are dancing. That's why the liquid in your thermometer rises on a hot day. Imagine the particles inside are dancing energetically, causing the liquid to expand.

Sometimes, adding or removing heat leads to a physical change in matter. When the ice melts in your glass of lemonade, you're witnessing a phase change. In this case, matter is changing from a solid (ice) to a liquid (water)!

You'll learn more about expansion and phase changes in the coming pages. Physicists study these topics so they can understand how different materials behave at different temperatures. Their research helps engineers build stable bridges, efficient rocket engines, and more!

Expand and **contract**

Holding a jar under hot running water warms it up. As temperature rises, ***matter*** expands. In the case of the stubborn jar lid, both the glass jar and the metal lid expand as they warm up. But the metal lid expands more, which loosens its tight grip on the jar.

A balloon left outside on a hot, sunny day will expand in all directions as the air inside it warms up. It may expand so much that it pops!

Have you ever wondered why a concrete sidewalk has built-in cracks? It's because even a material as hard as concrete expands due to temperature. Builders pour concrete in sections separated by spaces. The spaces allow room for the concrete to expand on hot days.

Next time you have a hard time opening a glass jar, try running hot water over the lid. That should make it easier to twist off the lid. Why? Let's find out!

Even structures as big as bridges expand. Have you ever noticed the seam where a bridge connects to the road leading up to it? This is called an expansion joint. These metal joints move together as the bridge expands in warmer weather and apart as the bridge contracts in cooler weather. Without these joints, an expanding bridge would press against the road and crack!

Changing **phases**

During a phase change, the addition or removal of heat causes an object's particles to rearrange. This changes the physical form of the object. Matter can take the form of a solid, liquid, or gas.

Particles in liquids are loosely packed and able to flow around each other. This particle motion gives liquids an indefinite shape. That's why liquids fit the container they are in.

Removing heat from a gas turns it into a liquid. When a cold drink cools the air around it, water vapor in the air loses energy. This causes the vapor to condense, or turn to liquid, on the cold glass.

Adding heat to a liquid changes it to a gas. For example, water's temperature rises until it reaches 212 °F (100 °C), which is the boiling point of water. After that, any additional heat will make the water change into steam, a gas.

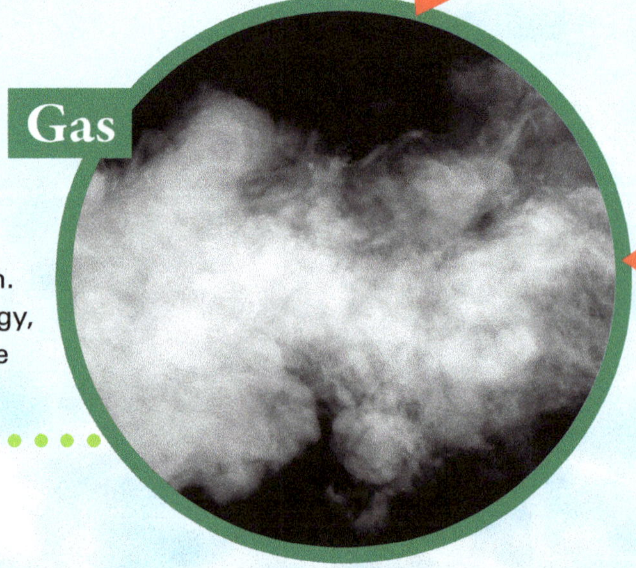

Particles in a gas bounce around with a lot of space between them. Because of this high kinetic energy, gas doesn't have a definite shape or volume.

Water, ice, and steam are different forms, or phases, of the same *matter*. So, what exactly happens during a phase change?

Adding heat to a solid turns it into a liquid. This happens when ice melts into water.

When a liquid loses energy, it becomes solid. This happens when water freezes into ice.

The particles in a solid are packed tightly together and don't move much. Because of this low kinetic energy, solids keep their shape well.

DID YOU KNOW?

Water is the only substance on Earth that occurs naturally as a solid, liquid, and gas.

5

EXTREME HOT AND COLD

Outer space is home to planets, moons, and stars. It's also home to the hottest and coldest temperatures known to scientists!

A cloud of gas and dust known as the Boomerang *Nebula* lies about 5,000 *light-years* from Earth. With a temperature of –457.87 °F (–272.15 °C), it is the coldest known place in the universe. It is, however, one degree warmer than absolute zero. Absolute zero is the lowest temperature anything can reach. At absolute zero, everything would be frozen solid, even the air we breathe! So far, no known substance has cooled all the way down to absolute zero.

High temperatures, on the other hand, have no limit. There is no such thing as absolute hot. In theory, an object can continue to get hotter and hotter and hotter. The hottest known place in the universe is the ***quasar*** 3C 273. This region about 2 billion light-years from Earth reaches 18 trillion °F (10 trillion °C)!

Back on Earth, the exhaust from the SpaceX Falcon 9 rocket is about 5500 °F (3000 °C). That's hot enough to melt metal!

Scientists are hard at work trying to reach the extremes of temperature. They know that surprising things happen on Earth and beyond when temperatures get as high and as low as they can go. Let's explore more about the extremes!

Absolute **zero**

As temperature decreases, particles lose kinetic energy and move more slowly. Absolute zero is the temperature at which ***atoms*** have no kinetic energy and are therefore almost motionless. The atoms still wiggle slightly, but they stay clumped together and move as one—kind of like one big superatom! This is a form of matter known as the Bose-Einstein condensate (BEC). Scientists are using BEC's to create new technologies and better understand our world.

In 2021, German scientists used lab equipment to reach 38 trillionths of a degree above 0 Kelvin. At the time, that was the closest scientists had gotten to absolute zero!

Absolute zero is 0 Kelvin, or −459.67 °F (−273.15 °C). Supercool things happen as ***matter*** approaches this limit!

Supercold matter acts in strange and potentially useful ways. For example, electricity loses some energy when the material it's traveling through creates ***resistance***. But superconductors are materials that can transmit electricity without any resistance or loss of energy. This leads to new technologies, including precision clocks, high-speed trains, and superfast computers. The problem is, superconductors function only at very low temperatures—that is, hundreds of degrees below the freezing point of water!

TECH TIME

The fastest trains in the world are known as maglev trains. That's short for "magnetic levitation." Superconducting magnets cooled to extremely low temperatures keep the train floating above the track for a smooth, fast ride.

A quantum processor is a superfast computer. It is surrounded by a chandelierlike structure that keeps the processor at temperatures close to absolute zero!

Heat **balance**

Much of the sun's radiation is *absorbed* or *reflected* by water vapor, dust, and other substances in Earth's atmosphere. Some radiation that reaches Earth is reflected off snow, ice, and other surfaces. The rest of the radiation is absorbed by Earth as thermal energy. This energy is reradiated from Earth's surface back to the atmosphere and to space.

Energy from the sun, called radiation, is responsible for heating Earth. In turn, Earth *radiates* energy back into space! A balance of incoming and outgoing heat is what keeps our planet livable.

What happens to the thermal energy that is reradiated from Earth's surface? Much of it gets trapped by water vapor, carbon dioxide, and other gases in the atmosphere. These gases act like a blanket that keeps Earth warm. Many human activities, such as burning fossil fuels, produce these heat-trapping gases. This is causing global warming, a long-term rise in Earth's temperature. Global warming has many negative effects, including more frequent droughts, wildfires, and severe weather.

TECH TIME

Researchers have designed devices that can send heat away from Earth. They look kind of like solar panels. But instead of collecting energy from the sun, these panels radiate energy into outer space. One day, this technology may be used to cool buildings without using electricity and help reduce global warming.

Hot sun, **cold space**

The sun's radiation travels through space, but it doesn't heat up space. This is because there is almost no ***matter*** in space. As a result, heat has nothing to transfer to, and space remains cold!

Radiation from the sun warms the surface of the moon—but not its entire surface at once! The side of the moon facing the sun measures up to 260 °F (127 °C). But the side of the moon facing away from the sun dips as low as –279 °F (–173 °C)!

The core of the sun is 27 million °F (15 million °C). Meanwhile, the average temperature of outer space is −454 °F (−270 °C). Why is space icy cold when the sun and other stars are so hot?

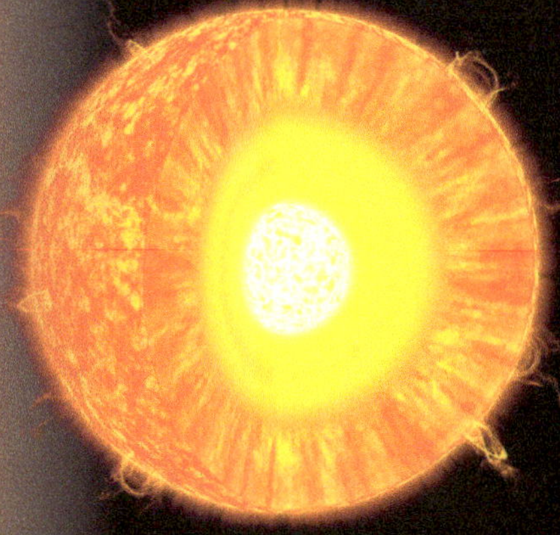

The universe began 13.8 billion years ago when a single hot, ***dense*** point began to expand at an incredible rate. Scientists call this event the big bang. Heat from the big bang continues to keep space above absolute zero!

TECH TIME

Astronauts wear special space suits to regulate their temperature during space walks. Tubes of chilled water keep the astronauts from overheating in their suits. Meanwhile, the gloves of the space suits have heaters to keep the astronauts' fingers warm!

Explore heat **transfer!**

You will need:

- A tall, clear glass or vase
- A smaller clear glass
- Hot (but not boiling) water
- Cold water
- Food coloring

Give it a try

1. Fill the tall glass or vase two-thirds full with cold water.
2. Fill the smaller glass with hot water. Add a few drops of food coloring so you can keep track of where the hot water is.
3. Carefully place the small glass into the tall container with the cold water. Then observe!
4. What happens to the colored hot water? What happens to the cold water?
5. What type of heat transfer did you demonstrate? How do you know?

Try this next!

What might happen if you use cold water in both glasses? Repeat the experiment, but this time, fill the smaller glass with cold water before you dye it with the food coloring. Watch what happens after you complete step 3. How does this result compare with your first experiment? Why might this be?

You don't have to bob in the sea to experience how heat moves around the ocean. You can use common household items to demonstrate heat transfer in tap water. Let's see how it works!

QUESTION TIME!

How do your experiments demonstrate why water flows from Earth's equator to the poles?

Index

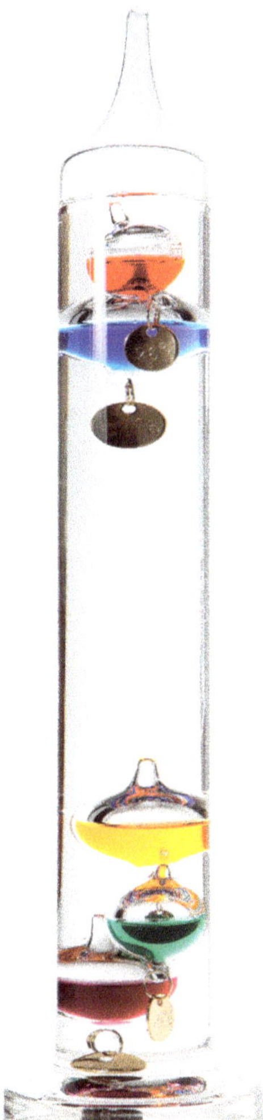

Glossary

absorb (ubb ZORB)—to soak up

atom (AT um)—the smallest unit of matter

dense (DENTS)—having a high mass per unit of volume; density is the measure of mass per unit of volume

friction (FRIK shuhn)—the force that slows or stops motion when two objects come into contact with each other

light-year (LITE yeer)—a unit of length equal to the distance that light travels in one year

mass (MAS)—a measure of the amount of matter in an object

matter (MAE tuhr)—the material that makes up physical objects

molecule (MAH lih kyool)—two or more atoms bonded together

nebula (NEB yoo luh)—a cloud of dust and gas where new stars may be born

quasar (KWAY zahr)—the superbright center of a galaxy that is extremely far away, likely produced by an active black hole

radiate (RAY dee ate)—to emit or spread rays from a central point

reflect (rih FLEKT)—to bounce or cause to bounce off an object

resistance (ree ZIH stents)—an opposing force

thermal equilibrium (THUR mul ee kwuh LIH bree uhm)—a state in which all parts of a system are at the same temperature

www.ingramcontent.com/pod-product-compliance
Lightning Source LLC
LaVergne TN
LVHW070858210926
848482LV00016B/969
9780716671633